For Kids Age 5–8

Vol. 1
Easy

This Book Belongs to:

A Very Brief History About Sudoku

Sudoku (数独) is a Japanese word; but its definitely based on an American game. Its origin most likely started in the 1780s with the Swiss Mathematician Leonhard Euler who developed the mathematical concept called "Latin Squares" which he often referred to as a 'new kind of magic squares'. The point of the game is to arrange numbers in a way that any number or symbol would occur only once in each row or column.

Howard Garns, an architect from Indianapolis, is credited for creating the rule that restrained each region to having only one number or symbol. Garns created "Number Place" and published it in *Dell Puzzle Magazine* in 1979 for over 25 years.

"Number Place" is translated as "Suuji wa dokushin ni kagiru" in Japanese, meaning the numbers must be 'unmarried or single'. The game was introduced to Japan by Publisher Nikoli in the paper *Monthly Nikolist* in April 1984. Japan added two rules: (1) the puzzle has to be symmetrical and (2) that 32 of 81 squares must be revealed to allow for some level of difficulty. The long name was later shortened to 'Sudoku' and remained in Japan for over 20 years.

Retired Hong Kong Judge Wayne Gould came across a Sudoku book in Tokyo in 1997 and dedicated 6 years to develop a computer program to make the puzzles. Gould approached *The Times* in Britain; and his game was launched on November 12, 2004, as "Su Doku". Gould published the game in United States with *The Conway Daily Sun* in 2004 as well.

Sudoku then became and is still an international craze.

Calcudoku, **Mathdoku**, or **Square Wisdom** is a mathematical and logical puzzle loosely similar to Sudoku. It was invented by Japanese mathematics teacher Tetsuya Miyamoto.

Sudoku & Brain Health

Brain games and puzzles, like Sudoku, help delay dementia & Alzheimer's disease as well as help the brain assess memory, attention & reasoning. Sudoku keeps the brain stimulated and can be very therapeutic.

How to Play

The objective is to fill the grid in with the digits 1 through N (where N is the number of rows or columns in the grid):

- Each row contains exactly one of each digit.

- Each column contains exactly one of each digit.

- Each **bold-outlined group** of cells contains digits which achieve the specified result using the specified mathematical operation: addition (+), subtraction (-), multiplication (×), and division (÷).

Digits may be repeated within a block.

Puzzle No: 1

6+	11+	3+	
7+			1−
6+			

Puzzle No: 2

2−		2−	
6+	3−		2−
	2−	9+	

Puzzle No: 3

3+	4+		1−
	8+		
10+		3−	1−

Puzzle No: 4

3−	1−		7+
	7+		
		9+	
5+			

Puzzle No: 5

2−	10+	6+	
		2−	3−
2−			
		1−	

Puzzle No: 6

8+	5+		8+
		2−	
	8+		
		1−	

Puzzle No: 7

11+		5+	2−
3+		9+	
8+			

Puzzle No: 8

6+		4+	
3−		1−	
1−	1−		3−
	7+		

Puzzle No: 9

11+			1−
1−	2−		
		2−	3−
2−			

Puzzle No: 10

9+	6+		
		1−	3−
3+	3−		
		1−	

Puzzle No: 11

8+	9+		
	5+	1−	
			2−
7+			

Puzzle No: 12

3−	2−	5+	
		2−	
1−		5+	4+
2−			

Puzzle No: 13

1−		8+	
9+			
11+		4+	5+

Puzzle No: 14

11+	2−		5+
1−		12+	
5+			

Puzzle No: 15

3+	10+		2−
		7+	
5+			2−
3−			

Puzzle No: 16

6+		5+	1−
	3−		
2−		8+	3−

Puzzle No: 17

2−	6+		3−
	1−		
2−		7+	
	7+		

Puzzle No: 18

2−	1−		4+
	6+		
		1−	
4+		2−	

Puzzle No: 19

7+		6+	
	2−		3−
6+	6+		
		2−	

Puzzle No: 20

6+			2−
5+		8+	
8+			
	7+		

Puzzle No: 21

2− 6+ 6+ 6+ 1− 2− 7+

Puzzle No: 22

1− 12+ 4+ 2− 10+ 1−

Puzzle No: 23

3− 7+ 1− 7+ 5+ 8+ 1−

Puzzle No: 24

6+ 1− 9+ 3+ 6+ 9+

Puzzle No: 25

6+	8+	2−	
			2−
6+	1−		
		1−	

Puzzle No: 26

9+			1−
11+			
	2−		3−
	6+		

Puzzle No: 27

4+	5+	2−	
		5+	2−
2−			
1−		5+	

Puzzle No: 28

6+			3−
2−	5+		
	8+		
5+		1−	

Puzzle No: 29

12+			8+
3+			
11+	3+		
		3+	

Puzzle No: 30

3−		5+	
12+	1−		4+
		9+	

Puzzle No: 31

4+		1−	9+
2−			
1−	11+		
		4+	

Puzzle No: 32

2−		6+	
1−		5+	3−
7+	6+		
		2−	

Puzzle No: 33

7+		6+	
1−	3−		9+
		3−	
5+			

Puzzle No: 34

9+		3−	
4+		1−	
	7+	4+	6+

Puzzle No: 35

2−	1−		3+
	6+		
8+			1−
	1−		

Puzzle No: 36

2−		6+	6+
7+			
		1−	
2−		4+	

Puzzle No: 37

3+ 8+ 2−
1−
1− 6+
3−

Puzzle No: 38

2− 10+
7+ 9+
5+
3−

Puzzle No: 39

3+ 9+ 3−
1−
1− 8+
3+

Puzzle No: 40

5+ 3−
2− 10+
6+ 3+
7+

Puzzle No: 41

6+	6+		
	8+		
3−		7+	
8+			

Puzzle No: 42

7+		11+	5+
6+			
	6+		3−

Puzzle No: 43

11+	5+		8+
2−		7+	
	3+		

Puzzle No: 44

1−		8+	
1−			1−
1−	12+		

Puzzle No: 45

10+ 7+ 5+ 7+ 6+ 1−

Puzzle No: 46

8+ 8+ 6+ 4+ 5+ 9+

Puzzle No: 47

6+ 7+ 10+ 3− 1− 5+

Puzzle No: 48

6+ 7+ 9+ 7+ 1− 1− 3−

Puzzle No: 49

3+	8+		
	8+		2−
7+			
3−		5+	

Puzzle No: 50

5+		3−	
11+		1−	
1−		8+	1−

Puzzle No: 51

10+			
6+		4+	
2−	6+		3+
	1−		

Puzzle No: 52

4+		7+	2−
2−	3+		
		2−	
10+			

Puzzle No: 53

3+	2−	7+	
		2−	
9+	2−		5+

Puzzle No: 54

8+		6+	
	2−	5+	
1−			1−
	7+		

Puzzle No: 55

3−	7+		6+
		3+	
9+			4+
	6+		

Puzzle No: 56

2−	7+		2−
	7+		
		2−	
3+		1−	

Puzzle No: 57

1−	3−		2−
	11+		
3−	4+		
		6+	

Puzzle No: 58

6+		11+	2−
7+			2−
	6+		

Puzzle No: 59

1−		5+	
2−	5+		1−
	2−		
2−		4+	

Puzzle No: 60

2−	9+		
	10+		
3−			3+
9+			

Puzzle No: 61

9+			3−
3−	5+	3+	
			1−
8+			

Puzzle No: 62

2−	3−		11+
1−		4+	3−
2−			

Puzzle No: 63

7+	1−		2−
		1−	
1−	1−		7+

Puzzle No: 64

7+		6+	3−
	13+		
		2−	
		1−	

Puzzle No: 65

11+		10+	
			6+
	1−		
10+			

Puzzle No: 66

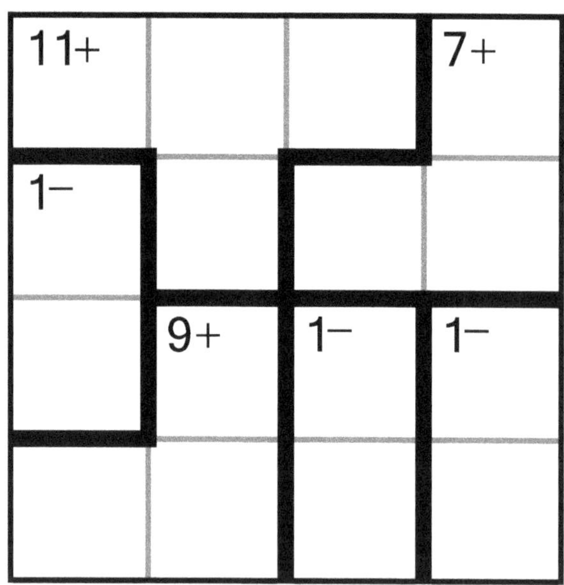

Puzzle No: 67

6+		12+	
1−			
	11+	5+	1−

Puzzle No: 68

3−		5+	
6+	3−		7+
		1−	
5+			

Puzzle No: 69

8+		14+	1−
3−	1−		
		1−	

Puzzle No: 70

1−		1−	7+
8+			
1−		4+	
		2−	

Puzzle No: 71

6+	2−		2−
	7+		
		2−	4+
1−			

Puzzle No: 72

6+		2−	
8+	1−		3+
	1−	9+	

Puzzle No: 73

2−	1−	2−	
		5+	3−
11+			
		6+	

Puzzle No: 74

6+			6+
1−	7+		
		11+	
1−			

Puzzle No: 75

1−		8+	10+
3−			
	5+		
1−			

Puzzle No: 76

2−	7+		6+
	5+	2−	
1−		5+	

Puzzle No: 77

9+		8+	
	2−	2−	
5+			8+

Puzzle No: 78

3−		8+	
6+	5+		
	10+	2−	

Puzzle No: 79

1−		6+	
10+		1−	
	1−		9+

Puzzle No: 80

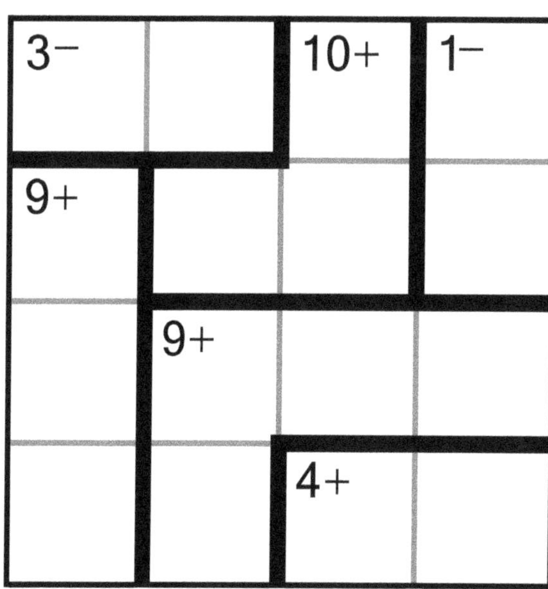

Puzzle No: 81

8+		1−	
1−		5+	7+
	9+		
		3−	

Puzzle No: 82

9+	7+		7+
		1−	
	8+		6+

Puzzle No: 83

7+	9+		
		2−	
2−		3+	6+
1−			

Puzzle No: 84

1−	2−	3−	
		9+	
7+			2−
	2−		

Puzzle No: 85

9+		9+	
			1−
2−	8+		
		3−	

Puzzle No: 86

1−		10+	
8+	2−		
		3−	1−
1−			

Puzzle No: 87

7+	2−		6+
	1−		
	8+		
1−		1−	

Puzzle No: 88

9+	7+		
	7+	2−	
		3−	1−
3+			

Puzzle No: 89

1−	1−		7+
	3−		
3−	1−	2−	
		4+	

Puzzle No: 90

6+			1−
3−	6+		
		6+	
7+		3+	

Puzzle No: 91

5+	6+		2−
	6+	3−	
3−			9+

Puzzle No: 92

1−	7+		
	5+		4+
1−	2−	8+	

Puzzle No: 93

5+		8+	
7+			9+
	7+		
2−			

Puzzle No: 94

4+	1−	2−	
		1−	2−
13+			
		3+	

Puzzle No: 95

9+		7+	5+
	2−		
		3−	
1−		1−	

Puzzle No: 96

3−		7+	
6+	2−		
	1−		3−
6+			

Puzzle No: 97

8+	1−		1−
	5+		
	2−		8+
1−			

Puzzle No: 98

4+		2−	
1−		1−	5+
3+	6+		
		2−	

Puzzle No: 99

1−		10+	3−
2−			
3−	4+		
		5+	

Puzzle No: 100

9+		1−	3+
	1−		
4+		3−	
	9+		

Puzzle No: 101

3−		8+	
9+	2−	1−	
			3−
	2−		

Puzzle No: 102

6+	9+	5+	
1−		13+	
2−			

Puzzle No: 103

2−	1−	11+	
		2−	
2−			
8+			

Puzzle No: 104

2−	5+	2−	
		7+	1−
4+	3−		
		2−	

Puzzle No: 105

11+		1−	1−
6+			
	3+		1−
	3−		

Puzzle No: 106

1−		2−	1−
6+			
6+		9+	
	3−		

Puzzle No: 107

9+		3−	
9+		1−	
		7+	
	1−		

Puzzle No: 108

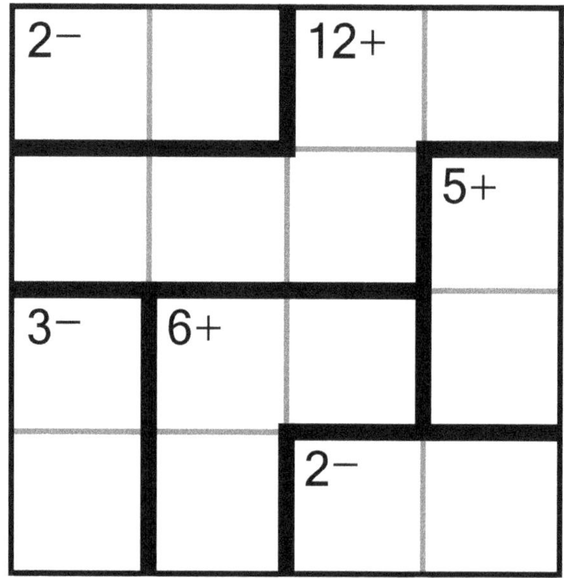

Puzzle No: 109

1−		7+	3−
1−			
	13+		
1−			

Puzzle No: 110

1−	2−		3+
	5+		
1−	1−		11+

Puzzle No: 111

3+		1−	
2−		5+	
8+			2−
	1−		

Puzzle No: 112

11+		1−	
	1−	2−	
4+		1−	6+

Puzzle No: 113

2−	2−	1−	
		6+	7+
2−			
	8+		

Puzzle No: 114

8+	3+		1−
	6+		
	6+		
1−		3−	

Puzzle No: 115

9+	3−		7+
	2−		
		3−	
1−		1−	

Puzzle No: 116

9+		2−	3−
2−			
	5+	2−	5+

Puzzle No: 117

10+			
6+	1−		3+
	2−		
2−		6+	

Puzzle No: 118

3−		9+	
1−	2−		
	7+		
7+		3+	

Puzzle No: 119

1−		1−	
5+		1−	
	1−		5+
6+			

Puzzle No: 120

6+	3+		2−
	10+	8+	
			2−
1−			

Puzzle No: 121

7+	2−		1−
1−	3−	5+	
		1−	

Puzzle No: 122

3−		1−	10+
4+			
1−	3−		
	6+		

Puzzle No: 123

10+		2−	1−
	6+		
		6+	7+
2−			

Puzzle No: 124

8+	1−		2−
		2−	
6+			2−
	7+		

Puzzle No: 125

2−		5+	
1−		6+	
2−	8+		6+

Puzzle No: 126

1−	8+		
	13+	3+	7+
		2−	

Puzzle No: 127

9+		3−	6+
2−			
	8+		
		2−	

Puzzle No: 128

8+		1−	7+
7+			
	3−	10+	

Puzzle No: 129

6+		8+	
	2−		
9+	7+		
		4+	

Puzzle No: 130

2−		7+	
8+	7+		
		4+	
5+		1−	

Puzzle No: 131

1−		8+	
7+			3+
3−	8+		
		2−	

Puzzle No: 132

1−		10+	5+
7+			
		1−	
1−		1−	

Puzzle No: 133

13+		2−	
	7+		
	5+		1−
8+			

Puzzle No: 134

7+	7+	4+	
		1−	
	13+		2−

Puzzle No: 135

8+			1−
13+	1−	6+	
			7+

Puzzle No: 136

2−	2−		10+
	1−		
6+	11+		

Puzzle No: 137

3−	2−		1−
	8+	8+	
			3−
1−			

Puzzle No: 138

9+	5+		10+
	7+	2−	
		1−	

Puzzle No: 139

2−	1−	4+	
		9+	
2−			1−
1−			

Puzzle No: 140

9+	1−		1−
		3+	
2−			11+
1−			

Puzzle No: 141

2−	6+	3−	
		1−	
2−	3−		2−
	5+		

Puzzle No: 142

3−	12+	7+	
2−		6+	
6+			

Puzzle No: 143

1−	4+		7+
	9+		
		10+	
3+			

Puzzle No: 144

2−		3−	7+
1−			
3−	5+		
	9+		

Puzzle No: 1

4	3	1	2
2	4	3	1
1	2	4	3
3	1	2	4

Puzzle No: 2

4	2	3	1
3	4	1	2
1	3	2	4
2	1	4	3

Puzzle No: 3

2	1	3	4
1	4	2	3
3	2	4	1
4	3	1	2

Puzzle No: 4

1	3	4	2
4	1	2	3
2	4	3	1
3	2	1	4

Puzzle No: 5

1	3	4	2
3	2	1	4
2	4	3	1
4	1	2	3

Puzzle No: 6

2	4	1	3
3	2	4	1
1	3	2	4
4	1	3	2

Puzzle No: 7

3	4	2	1
4	2	1	3
2	1	3	4
1	3	4	2

Puzzle No: 8

4	2	1	3
1	4	3	2
3	1	2	4
2	3	4	1

Puzzle No: 9

1	4	2	3
3	1	4	2
2	3	1	4
4	2	3	1

Puzzle No: 10

4	3	1	2
3	2	4	1
2	1	3	4
1	4	2	3

Puzzle No: 11

3	1	2	4
1	3	4	2
4	2	3	1
2	4	1	3

Puzzle No: 12

4	1	3	2
1	3	2	4
3	2	4	1
2	4	1	3

Puzzle No: 13

1	2	3	4
2	3	4	1
3	4	1	2
4	1	2	3

Puzzle No: 14

4	1	3	2
3	4	2	1
1	2	4	3
2	3	1	4

Puzzle No: 15

1	4	3	2
2	3	1	4
3	2	4	1
4	1	2	3

Puzzle No: 16

1	2	4	3
3	4	1	2
2	1	3	4
4	3	2	1

Puzzle No: 17

2	1	3	4
4	3	2	1
1	2	4	3
3	4	1	2

Puzzle No: 18

2	4	3	1
4	3	1	2
1	2	4	3
3	1	2	4

Puzzle No: 19

1	3	4	2
3	4	2	1
2	1	3	4
4	2	1	3

Puzzle No: 20

3	1	2	4
1	4	3	2
4	2	1	3
2	3	4	1

Puzzle No: 21

3	1	2	4
4	3	1	2
2	4	3	1
1	2	4	3

Puzzle No: 22

3	4	2	1
2	3	1	4
4	1	3	2
1	2	4	3

Puzzle No: 23

1	4	2	3
4	3	1	2
3	2	4	1
2	1	3	4

Puzzle No: 24

1	4	3	2
2	1	4	3
3	2	1	4
4	3	2	1

Puzzle No: 25

2	4	3	1
4	3	1	2
3	1	2	4
1	2	4	3

Puzzle No: 26

1	3	4	2
4	2	1	3
2	1	3	4
3	4	2	1

Puzzle No: 27

3	1	4	2
1	4	2	3
4	2	3	1
2	3	1	4

Puzzle No: 28

3	2	1	4
4	3	2	1
2	1	4	3
1	4	3	2

Puzzle No: 29

2	3	4	1
1	2	3	4
4	1	2	3
3	4	1	2

Puzzle No: 30

1	4	3	2
4	3	2	1
2	1	4	3
3	2	1	4

Puzzle No: 31

3	1	2	4
4	2	1	3
1	3	4	2
2	4	3	1

Puzzle No: 32

1	3	4	2
2	1	3	4
3	4	2	1
4	2	1	3

Puzzle No: 33

4	3	2	1
1	4	3	2
2	1	4	3
3	2	1	4

Puzzle No: 34

2	3	4	1
1	4	2	3
3	2	1	4
4	1	3	2

Puzzle No: 35

2	3	4	1
4	1	3	2
1	4	2	3
3	2	1	4

Puzzle No: 36

3	1	2	4
4	3	1	2
1	2	4	3
2	4	3	1

Puzzle No: 37

2	3	1	4
1	4	3	2
3	2	4	1
4	1	2	3

Puzzle No: 38

3	1	2	4
4	2	3	1
1	3	4	2
2	4	1	3

Puzzle No: 39

2	3	1	4
1	2	4	3
4	1	3	2
3	4	2	1

Puzzle No: 40

3	2	1	4
1	3	4	2
4	1	2	3
2	4	3	1

Puzzle No: 41

4	2	3	1
2	3	1	4
1	4	2	3
3	1	4	2

Puzzle No: 42

4	3	1	2
1	4	2	3
3	2	4	1
2	1	3	4

Puzzle No: 43

3	1	4	2
1	3	2	4
2	4	3	1
4	2	1	3

Puzzle No: 44

1	2	3	4
4	3	1	2
2	1	4	3
3	4	2	1

Puzzle No: 45

1	2	4	3
4	1	3	2
3	4	2	1
2	3	1	4

Puzzle No: 46

4	2	3	1
2	3	1	4
3	1	4	2
1	4	2	3

Puzzle No: 47

2	1	3	4
3	2	4	1
4	3	1	2
1	4	2	3

Puzzle No: 48

1	2	3	4
2	4	1	3
4	3	2	1
3	1	4	2

Puzzle No: 49

2	3	1	4
1	2	4	3
3	4	2	1
4	1	3	2

Puzzle No: 50

3	2	1	4
4	3	2	1
1	4	3	2
2	1	4	3

Puzzle No: 51

2	1	3	4
4	2	1	3
3	4	2	1
1	3	4	2

Puzzle No: 52

1	3	4	2
2	1	3	4
4	2	1	3
3	4	2	1

Puzzle No: 53

2	1	3	4
1	3	4	2
3	4	2	1
4	2	1	3

Puzzle No: 54

3	1	2	4
4	2	3	1
2	4	1	3
1	3	4	2

Puzzle No: 55

4	1	3	2
1	3	2	4
2	4	1	3
3	2	4	1

Puzzle No: 56

2	4	1	3
4	3	2	1
3	1	4	2
1	2	3	4

Puzzle No: 57

2	4	1	3
3	2	4	1
4	1	3	2
1	3	2	4

Puzzle No: 58

3	1	4	2
2	3	1	4
4	2	3	1
1	4	2	3

Puzzle No: 59

2	3	1	4
3	1	4	2
1	4	2	3
4	2	3	1

Puzzle No: 60

1	4	2	3
3	2	1	4
4	1	3	2
2	3	4	1

Puzzle No: 61

2	4	3	1
1	3	2	4
4	2	1	3
3	1	4	2

Puzzle No: 62

3	1	4	2
1	4	2	3
2	3	1	4
4	2	3	1

Puzzle No: 63

2	3	4	1
1	4	2	3
4	1	3	2
3	2	1	4

Puzzle No: 64

1	3	2	4
3	2	4	1
2	4	1	3
4	1	3	2

Puzzle No: 65

2	1	4	3
1	4	3	2
3	2	1	4
4	3	2	1

Puzzle No: 66

4	1	3	2
2	3	4	1
1	4	2	3
3	2	1	4

Puzzle No: 67

3	2	1	4
2	1	4	3
1	4	3	2
4	3	2	1

Puzzle No: 68

4	1	2	3
3	4	1	2
1	2	3	4
2	3	4	1

Puzzle No: 69

2	3	4	1
3	4	1	2
4	1	2	3
1	2	3	4

Puzzle No: 70

2	3	1	4
1	4	2	3
4	2	3	1
3	1	4	2

Puzzle No: 71

2	1	3	4
3	4	1	2
1	2	4	3
4	3	2	1

Puzzle No: 72

2	4	1	3
1	3	4	2
4	2	3	1
3	1	2	4

Puzzle No: 73

4	2	1	3
2	1	3	4
3	4	2	1
1	3	4	2

Puzzle No: 74

1	3	2	4
3	4	1	2
4	2	3	1
2	1	4	3

Puzzle No: 75

3	4	2	1
1	2	4	3
4	1	3	2
2	3	1	4

Puzzle No: 76

2	4	3	1
4	1	2	3
1	3	4	2
3	2	1	4

Puzzle No: 77

2	3	4	1
4	2	1	3
1	4	3	2
3	1	2	4

Puzzle No: 78

4	1	3	2
2	4	1	3
1	3	2	4
3	2	4	1

Puzzle No: 79

4	3	1	2
1	4	2	3
2	1	3	4
3	2	4	1

Puzzle No: 80

1	4	3	2
2	3	4	1
3	1	2	4
4	2	1	3

Puzzle No: 81

4	3	1	2
2	1	3	4
1	4	2	3
3	2	4	1

Puzzle No: 82

2	1	4	3
3	2	1	4
4	3	2	1
1	4	3	2

Puzzle No: 83

1	2	4	3
2	4	3	1
3	1	2	4
4	3	1	2

Puzzle No: 84

2	3	1	4
3	1	4	2
4	2	3	1
1	4	2	3

Puzzle No: 85

1	4	2	3
3	1	4	2
4	2	3	1
2	3	1	4

Puzzle No: 86

2	1	3	4
1	4	2	3
4	3	1	2
3	2	4	1

Puzzle No: 87

4	3	1	2
1	2	3	4
2	1	4	3
3	4	2	1

Puzzle No: 88

3	1	2	4
2	4	3	1
4	3	1	2
1	2	4	3

Puzzle No: 89

3	1	2	4
2	4	1	3
1	3	4	2
4	2	3	1

Puzzle No: 90

2	1	3	4
4	2	1	3
1	3	4	2
3	4	2	1

Puzzle No: 91

3	4	2	1
2	1	4	3
4	3	1	2
1	2	3	4

Puzzle No: 92

3	1	4	2
4	3	2	1
2	4	1	3
1	2	3	4

Puzzle No: 93

3	2	4	1
4	1	3	2
2	4	1	3
1	3	2	4

Puzzle No: 94

3	1	2	4
1	2	4	3
2	4	3	1
4	3	1	2

Puzzle No: 95

1	2	4	3
4	1	3	2
2	3	1	4
3	4	2	1

Puzzle No: 96

1	4	2	3
4	1	3	2
2	3	4	1
3	2	1	4

Puzzle No: 97

4	2	3	1
3	1	4	2
1	4	2	3
2	3	1	4

Puzzle No: 98

3	1	4	2
4	3	2	1
1	2	3	4
2	4	1	3

Puzzle No: 99

3	2	1	4
2	4	3	1
1	3	4	2
4	1	2	3

Puzzle No: 100

2	3	4	1
4	1	3	2
3	2	1	4
1	4	2	3

Puzzle No: 101

1	4	3	2
4	1	2	3
2	3	1	4
3	2	4	1

Puzzle No: 102

2	4	1	3
4	3	2	1
1	2	3	4
3	1	4	2

Puzzle No: 103

4	2	1	3
2	3	4	1
3	1	2	4
1	4	3	2

Puzzle No: 104

4	2	1	3
2	3	4	1
1	4	3	2
3	1	2	4

Puzzle No: 105

4	3	2	1
1	4	3	2
3	2	1	4
2	1	4	3

Puzzle No: 106

3	4	1	2
4	2	3	1
1	3	2	4
2	1	4	3

Puzzle No: 107

2	3	1	4
1	4	2	3
3	1	4	2
4	2	3	1

Puzzle No: 108

2	4	3	1
3	1	4	2
4	2	1	3
1	3	2	4

Puzzle No: 109

2	3	1	4
3	4	2	1
4	1	3	2
1	2	4	3

Puzzle No: 110

3	4	2	1
4	1	3	2
2	3	1	4
1	2	4	3

Puzzle No: 111

2	1	4	3
4	2	3	1
3	4	1	2
1	3	2	4

Puzzle No: 112

3	4	2	1
4	2	1	3
1	3	4	2
2	1	3	4

Puzzle No: 113

3	4	1	2
1	2	3	4
4	1	2	3
2	3	4	1

Puzzle No: 114

3	2	1	4
1	4	2	3
4	1	3	2
2	3	4	1

Puzzle No: 115

3	4	1	2
4	1	2	3
2	3	4	1
1	2	3	4

Puzzle No: 116

3	2	1	4
2	4	3	1
4	1	2	3
1	3	4	2

Puzzle No: 117

1	2	4	3
2	4	3	1
4	3	1	2
3	1	2	4

Puzzle No: 118

1	4	2	3
2	1	3	4
3	2	4	1
4	3	1	2

Puzzle No: 119

4	3	1	2
1	2	4	3
2	4	3	1
3	1	2	4

Puzzle No: 120

4	1	2	3
2	3	4	1
3	4	1	2
1	2	3	4

Puzzle No: 121

1	2	4	3
2	3	1	4
4	1	3	2
3	4	2	1

Puzzle No: 122

4	1	3	2
1	3	2	4
2	4	1	3
3	2	4	1

Puzzle No: 123

2	4	3	1
4	3	1	2
1	2	4	3
3	1	2	4

Puzzle No: 124

3	2	1	4
1	3	4	2
4	1	2	3
2	4	3	1

Puzzle No: 125

4	2	1	3
2	3	4	1
3	1	2	4
1	4	3	2

Puzzle No: 126

2	4	3	1
1	3	2	4
4	2	1	3
3	1	4	2

Puzzle No: 127

3	2	4	1
2	4	1	3
4	1	3	2
1	3	2	4

Puzzle No: 128

3	2	1	4
4	3	2	1
1	4	3	2
2	1	4	3

Puzzle No: 129

2	3	1	4
1	4	2	3
3	1	4	2
4	2	3	1

Puzzle No: 130

3	1	2	4
2	3	4	1
4	2	1	3
1	4	3	2

Puzzle No: 131

2	1	4	3
3	4	1	2
4	2	3	1
1	3	2	4

Puzzle No: 132

1	2	3	4
2	3	4	1
4	1	2	3
3	4	1	2

Puzzle No: 133

2	4	1	3
3	1	2	4
4	2	3	1
1	3	4	2

Puzzle No: 134

2	4	1	3
4	3	2	1
1	2	3	4
3	1	4	2

Puzzle No: 135

1	3	4	2
4	2	3	1
3	1	2	4
2	4	1	3

Puzzle No: 136

2	3	1	4
4	1	2	3
1	4	3	2
3	2	4	1

Puzzle No: 137

1	4	2	3
4	3	1	2
3	2	4	1
2	1	3	4

Puzzle No: 138

2	3	1	4
3	1	4	2
4	2	3	1
1	4	2	3

Puzzle No: 139

4	3	1	2
2	4	3	1
3	1	2	4
1	2	4	3

Puzzle No: 140

2	4	3	1
4	3	1	2
3	1	2	4
1	2	4	3

Puzzle No: 141

3	2	1	4
1	4	3	2
2	1	4	3
4	3	2	1

Puzzle No: 142

1	2	4	3
4	3	1	2
3	1	2	4
2	4	3	1

Puzzle No: 143

3	1	2	4
4	3	1	2
2	4	3	1
1	2	4	3

Puzzle No: 144

3	1	4	2
2	3	1	4
4	2	3	1
1	4	2	3

This book is dedicated to my Grandfather, an avid Chinese chess strategist, who lovingly shared with me his wealth of Vietnamese riddles and alliterations when I was a very young child... but more importantly, who taught me that strategy games and riddles are wonderful pastimes to keep your mind active whether spent alone and better with loved ones.

Thank YOU for sharing your time with our puzzles.

To receive your BONUS puzzles, please send an email with your First Name and code CC-Kids01 to:
LudiYouXi@gmail.com

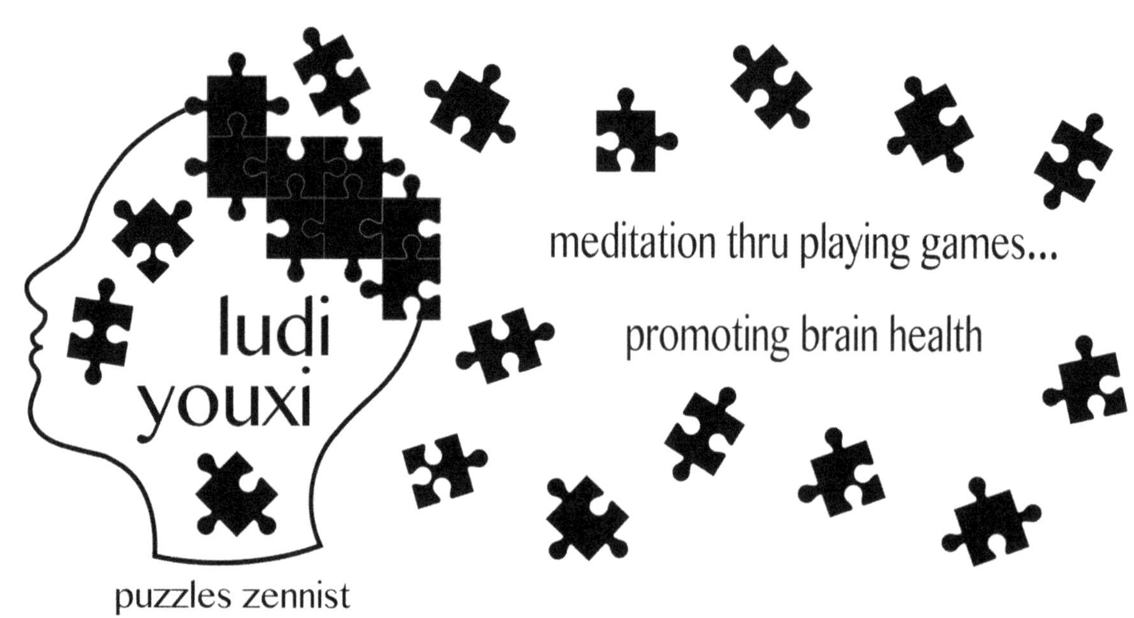

www.ingramcontent.com/pod-product-compliance
Ingram Content Group UK Ltd.
Pitfield, Milton Keynes, MK11 3LW, UK
UKHW061658200726
13853UKWH00013B/2503

9 798506 179856